Calculus 1 and One

Sebastien Simon

Copyright © 2017 Sebastien Simon

All rights reserved.

ISBN: 1542974216
ISBN-13: 978-1542974219

DISCLAIMER

The material in this book is a guide on how to do certain calculus 1 problems step-by-step. This book should not substitute for a textbook.

DEDICATION

Dedicated to anyone who love math.

CONTENTS

1	AN INTRODUCTION TO LIMITS	Pg 1
2	LEARN DIFFERENT WAY LIMITS CAN FAIL TO EXIST	Pg 2
3	DIVIDING & RATIONALIZE TECHNIQUE	Pg 3
4	INFINITE LIMITS	Pg 7
5	CONTINUITY AND ONE-SIDED LIMITS	Pg 8
6	LIMITS DEFINITION OF A DERIVATIVE	Pg 11
7	DERIVATIVE & TANGENT	Pg 13
8	DERIVATIVE	Pg 17
9	IMPLICIT DIFFERENTIATION	Pg 19
10	RELATED RATES	Pg 21

11	EXTREMA ON AN INTERVAL	Pg 26
12	INCREASING AND DECREASING FUNCTIONS AND THE FIRST DERIVATIVE TEST	Pg 27
13	CONCAVITY AND SECOND DERIVATIVE TEST	Pg 30
14	OPTIMIZATION	Pg 32
15	DIFFERENTIAL	Pg 35
16	ANTI-DERIVATIVES AND INDEFINITE INTEGRATION	Pg 36
17	INTEGRATION BY SUBSTITUTION	Pg 37
18	RIEMANN SUMS & DEFINITE INTEGRALS	Pg 40
19	BIBLIOGRAPHY	Pg 48

1 AN INTRODUCTION TO LIMITS

$$\lim_{x \to 3} \left(\frac{x^2 - 4}{x - 2} \right)$$

$$\lim_{x \to 0} (\tan x)$$

$$\lim_{x \to 3} \left(\frac{(x+2)(x-2)}{x-2} \right)$$

$$\lim_{x \to 0} \left(\frac{\sin(x)}{\cos(x)} \right)$$

$$\lim_{x \to 3} (x + 2)$$

$$\frac{\sin(0)}{\cos(0)} = \boxed{0}$$

$$(3 + 2) = \boxed{5}$$

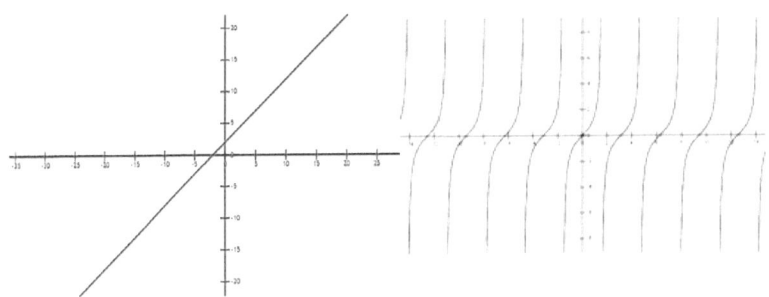

2 LEARN DIFFERENT WAY LIMITS CAN FAIL TO EXIST

1) Contradict itself
2) Unbounded behaviors
3) Oscillating behavior

$$\lim_{x \to 6}\left(\frac{3}{x-6}\right) \qquad \lim_{x \to 5}\left(\frac{|x-5|}{x-5}\right)$$

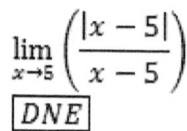

$$\frac{3}{0} = \boxed{DNE}$$

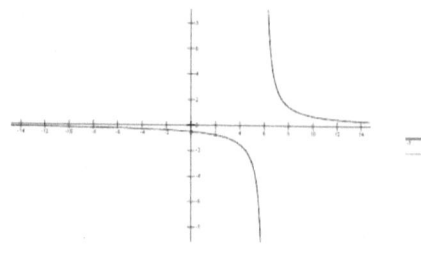
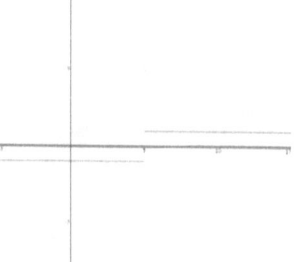

3 DIVIDING & RATIONALIZE TECHNIQUE

$$\lim_{x \to 0} \left(\frac{x^9 - 2x^3}{x^3} \right)$$

$$\lim_{x \to -1} \left(\frac{x^3 + 8}{x + 2} \right)$$

$$\lim_{x \to 0} \left(\frac{x^3(x^6 - 2)}{x^3} \right)$$

$$\lim_{x \to -1} \left(\frac{(x+2)(x^2 - 2x + 4)}{(x+2)} \right)$$

$$\lim_{x \to 0} (x^6 - 2)$$

$$\lim_{x \to -1} (x^2 - 2x + 4)$$

$((0)^6 - 2) = \boxed{-2}$

$(-1)^2 - 2(-1) + 4$
$\boxed{7}$

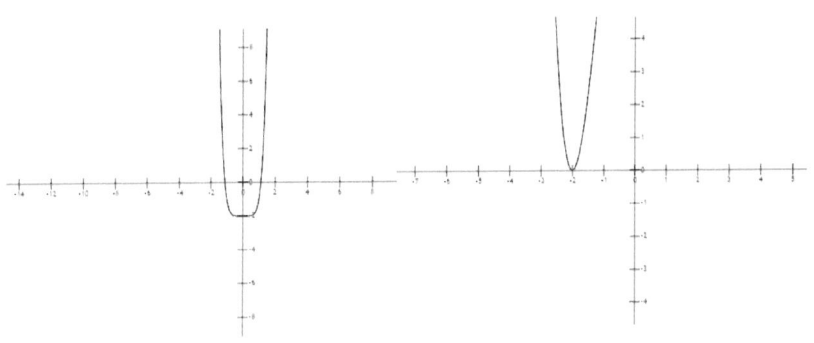

$$\lim_{x \to 3} \left(\frac{\sqrt{x+1} - 5}{x - 24} \right)$$

$$\lim_{x \to 3} \left(\frac{\sqrt{x+1} - 5}{x - 24} \right) \frac{(\sqrt{x+1} + 5)}{(\sqrt{x+1} + 5)}$$

$$\lim_{x \to 3} \left(\frac{(\sqrt{x+1})^2 + 5\sqrt{x+1} - 5\sqrt{x+1} - 25}{(x - 24)(\sqrt{x+1} + 5)} \right)$$

$$\lim_{x \to 3} \left(\frac{x + 1 - 25}{(x - 24)(\sqrt{x+1} + 5)} \right)$$

$$\lim_{x \to 3} \left(\frac{x - 24}{(x - 24)(\sqrt{x+1} + 5)} \right)$$

$$\lim_{x \to 3} \left(\frac{1}{\sqrt{x+1} + 5} \right)$$

$$\frac{1}{\sqrt{(3)+1} + 5}$$

$$\boxed{\frac{1}{7}}$$

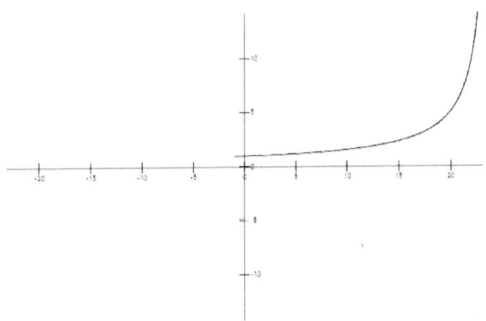

$$\lim_{x \to 0} \left(\left(\frac{\frac{1}{x+5} - \frac{1}{5}}{x} \right) \right)$$

$$\lim_{x \to 0} \left(\left(\frac{\frac{5}{5(x+5)} - \frac{1(x+5)}{5(x+5)}}{x} \right) \right)$$

$$\lim_{x \to 0} \left(\frac{5 - 5 - x}{5(x+5)} \right)$$

$$\lim_{x \to 0} \left(\frac{-x}{5(x+5)} \times \frac{1}{x} \right)$$

$$\lim_{x \to 0} \left(\frac{-1}{5(x+5)} \right)$$

$$\boxed{-\frac{1}{25}}$$

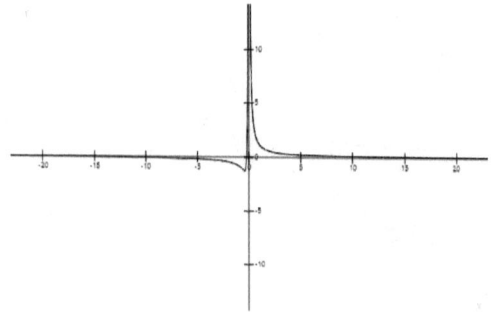

4 INFINITE LIMITS

$$\lim_{x \to 6^+} \left(\frac{-4}{x-6} \right)$$

$$\lim_{x \to 6^+} \left(\frac{-4}{x-6} \right) = -\infty$$

$$f(x) = -\frac{1}{(x-4)^2}$$

$$\lim_{x \to 6^-} \left(-\frac{1}{(x-6)^2} \right) = -\infty$$

$$\lim_{x \to 6^+} \left(-\frac{1}{(x-6)^2} \right) = -\infty$$

$$\lim_{x \to \infty} \left(\left(\frac{\frac{6}{x^3} - \frac{2}{x^3}}{\frac{2x^3}{x^3} + \frac{5x^2}{x^3} + \frac{5x}{x^3}} \right) \right)$$

$$\lim_{x \to \infty} \left(\left(\frac{\frac{6}{1} - \frac{2}{x^3}}{\frac{2}{1} + \frac{5}{x} + \frac{5}{x^2}} \right) \right)$$

3

5 CONTINUITY AND ONE-SIDED LIMITS

$$\lim_{x \to 4^-} \left(\frac{3}{x+2} \right) \qquad\qquad \lim_{x \to 3^+} \left(\frac{(x-5)}{(x-5)(x+5)} \right)$$

$$\lim_{x \to 4^-} \left(\frac{3}{4+2} \right) \qquad\qquad \lim_{x \to 3^+} \left(\frac{1}{x+5} \right)$$

$$\boxed{\frac{1}{2}} \qquad\qquad\qquad\qquad \boxed{\frac{1}{8}}$$

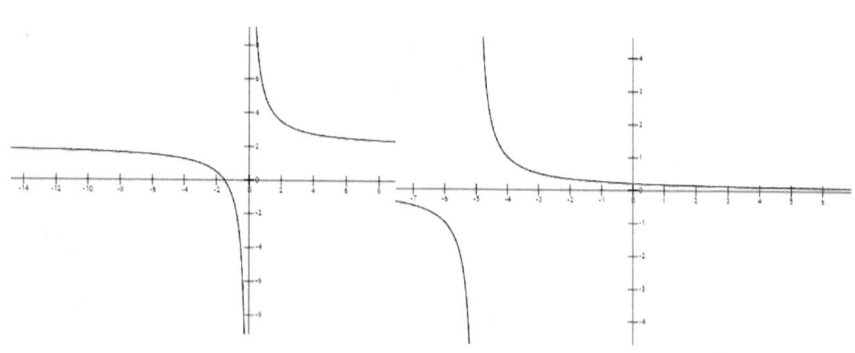

$$f(x) = \frac{x+3}{(x+3)(x-2)}$$

$x = -3, 2$

$x = -3$

$f(-3) = 0$
$$\lim_{x \to -3} \left(\frac{x+3}{(x+3)(x-2)} \right)$$
$$\lim_{x \to -3} \left(\frac{1}{x-2} \right) = -\frac{1}{5}$$
Removable, $x = -3$

$x = 2$
$$f(2) = \frac{5}{2} = Undefined$$
$$\lim_{x \to 2} \left(\frac{x+3}{(x+3)(x-2)} \right) = D.N.E$$

Nonremovable

$$f(x) = \begin{cases} -2x, & x \leq 2 \\ x^2 - 4x + 1, & x > 2 \end{cases}$$

$f(2) = -2(2) = -4$
$\lim_{x \to 2^+} (2^2 - 4(2) + 1) = 3$
$\lim_{x \to 2^-} (-2(2)) = -4$
$\lim_{x \to 2} (f(x)) = D.N.E$
Nonremovable $x = 2$

$$f(x) = \tan\frac{\pi x}{2}$$
$$\frac{\pi}{b} = period$$
$$\pi \div \frac{\pi}{2} = 2$$
Nonremovable $x = 1 + 2n$,
where $n = 0, 1, 2, 3 \dots$

6 LIMITS DEFINITION OF A DERIVATIVE

$$\lim_{\Delta x \to 0} \left(\frac{(x + \Delta x)^2 - x^2}{\Delta x} \right)$$

$$\lim_{\Delta x \to 0} \left(\frac{x^2 + 2x(\Delta x) + (\Delta x)^2 - x^2}{\Delta x} \right)$$

$$\lim_{\Delta x \to 0} \left(\frac{2x(\Delta x) + (\Delta x)^2}{\Delta x} \right)$$

$$\lim_{\Delta x \to 0} \left(\frac{\Delta x (2x + \Delta x)}{\Delta x} \right)$$

$$\lim_{\Delta x \to 0} (2x + \Delta x)$$

$$\boxed{2x}$$

$$\lim_{\Delta x \to 0} \left(\frac{f(x + \Delta x) - f(x)}{\Delta x} \right)$$

Given:
$f(x) = x^2 - 4x$

$$\lim_{\Delta x \to 0} \left(\frac{(x + \Delta x)^2 - 4(x + \Delta x) - (x^2 - 4x)}{\Delta x} \right)$$

$$\lim_{\Delta x \to 0} \left(\frac{x^2 + 2x(\Delta x) + (\Delta x)^2 - 4(x + \Delta x) - (x^2 - 4x)}{\Delta x} \right)$$

$$\lim_{\Delta x \to 0} \left(\frac{2x(\Delta x) + (\Delta x)^2 - 4(\Delta x)}{\Delta x} \right)$$

$$\lim_{\Delta x \to 0} \left(\frac{\Delta x (2x + \Delta x - 4)}{\Delta x} \right)$$

$$\lim_{\Delta x \to 0} (2x + \Delta x - 4)$$

$$\boxed{2x - 4}$$

7 DERIVATIVE & TANGENT

Find the slope of tan line of the graph given point
$f(x) = x^2 + 3x - 4$
$(-1, -6)$

$$\lim_{\Delta x \to 0} \left(\frac{(x + \Delta x)^2 + 3(x + \Delta x) - 4 - (x^2 + 3x - 4)}{\Delta x} \right)$$

$$\lim_{\Delta x \to 0} \left(\frac{x^2 + 2x\Delta x + (\Delta x)^2 + 3x + 3\Delta x - 4 - x^2 - 3x + 4}{\Delta x} \right)$$

$$\lim_{\Delta x \to 0} \left(\frac{2x\Delta x + (\Delta x)^2 + 3\Delta x}{\Delta x} \right)$$

$$\lim_{\Delta x \to 0} \left(\frac{\Delta x(2x + \Delta x + 3)}{\Delta x} \right)$$

$$\lim_{\Delta x \to 0} (2x + \Delta x + 3) = 2x + 3$$

$m_{tan}(-1) = 2(-1) + 3$

1

Find the equation of the tangent line to the graph of the equation at the given point
$f(x) = x^2 + 3x - 4$
$(-1, -6)$

$$\lim_{\Delta x \to 0} \left(\frac{(x + \Delta x)^2 + 3(x + \Delta x) - 4 - (x^2 + 3x - 4)}{\Delta x} \right)$$

$$\lim_{\Delta x \to 0} \left(\frac{x^2 + 2x\Delta x + (\Delta x)^2 + 3x + 3\Delta x - 4 - x^2 - 3x + 4}{\Delta x} \right)$$

$$\lim_{\Delta x \to 0} \left(\frac{2x\Delta x + (\Delta x)^2 + 3\Delta x}{\Delta x} \right)$$

$$\lim_{\Delta x \to 0} \left(\frac{\Delta x(2x + \Delta x + 3)}{\Delta x} \right)$$

$\lim_{\Delta x \to 0} (2x + \Delta x + 3) = 2x + 3$
$2(-1) + 3 = 1$

$y + 6 = 1(x + 1)$

$\boxed{y = x - 5}$

$$f(x) = \frac{-5}{\sqrt{x}}$$

$$\lim_{\Delta x \to 0} \left(\frac{\frac{1}{\sqrt{x+\Delta x}} - \frac{1}{\sqrt{x}}}{\Delta x} \right)$$

$$-5 \lim_{\Delta x \to 0} \left(\frac{\frac{\sqrt{x}}{\sqrt{x}} - \frac{\sqrt{x+\Delta x}}{\sqrt{x+\Delta x}}}{\frac{\Delta x}{1}} \right)$$

$$-5 \lim_{\Delta x \to 0} \left(\frac{\sqrt{x} - \sqrt{x+\Delta x}}{\sqrt{x}\sqrt{x+\Delta x}} \cdot \frac{1}{\Delta x} \right)$$

$$-5 \lim_{\Delta x \to 0} \left(\frac{\sqrt{x} - \sqrt{x+\Delta x}}{\Delta x \sqrt{x}\sqrt{x+\Delta x}} \cdot \frac{\sqrt{x} + \sqrt{x+\Delta x}}{\sqrt{x} + \sqrt{x+\Delta x}} \right)$$

$$-5 \lim_{\Delta x \to 0} \left(\frac{x - (x+\Delta x)}{\Delta x \sqrt{x}\sqrt{x+\Delta x}(\sqrt{x} + \sqrt{x+\Delta x})} \right)$$

$$-5 \lim_{\Delta x \to 0} \left(\frac{-(\Delta x)}{\Delta x \sqrt{x}\sqrt{x+\Delta x}(\sqrt{x} + \sqrt{x+\Delta x})} \right)$$

$$-5 \lim_{\Delta x \to 0} \left(\frac{-1}{\sqrt{x}\sqrt{x+\Delta x}(\sqrt{x}+\sqrt{x+\Delta x})} \right)$$

$$-5 \left[\frac{-1}{x(2\sqrt{x})} \right]$$

$$\boxed{\frac{5}{2x\sqrt{x}}}$$

8 DERIVATIVE

$f(x) = 5x^4$

$5[4x^{4-1}]$

$\boxed{f'(x) = 20x^3}$

$f(x) = -2\sqrt[3]{x^2}$

$-2\dfrac{d}{dx}\left(\sqrt[3]{x^2}\right)$

Let $(x^2) = u$

$-2\dfrac{d}{du}\left(\sqrt[3]{u}\right)\dfrac{d}{dx}(x^2)$

$\dfrac{d}{du}\left(\sqrt[3]{u}\right) = \dfrac{1}{3u^{\frac{2}{3}}}$

$\dfrac{d}{dx}(x^2) = 2x$

$-2\left(\dfrac{1}{3u^{\frac{2}{3}}}\right) \times 2x$

$-2\left(\dfrac{1}{3(x^2)^{\frac{2}{3}}}\right) \times 2x$

$\boxed{f'(x) = \left(\dfrac{-4x}{3(x^2)^{\frac{2}{3}}}\right)}$

$f(x) = \dfrac{4}{x^3}$

$-4[-3x^{-3-1}]$

$\boxed{f'(x) = \dfrac{12}{x^4}}$

$f(x) = \dfrac{\cos x}{\sin x}$

$\dfrac{\sin x(-\sin x) - \cos x(\cos x)}{\sin^2 x}$

$\dfrac{-\sin^2 x - \cos^2 x}{\sin^2 x}$

$\dfrac{1}{\sin^2 x}$

$$\boxed{f'(x) = -\csc^2 x}$$

$$\boxed{f'(x) = \frac{1 - 5\ln x}{x^6}}$$

$f(x) = \sec x$

$$\sec x = \frac{1}{\cos x}$$

$$\frac{\cos x(0) - 1(-\sin x)}{\cos^2 x}$$

$$\frac{\sin x}{\cos^2 x}$$

$$\boxed{f'(x) = \sec x \tan x}$$

$$\frac{d}{dx}\left[\frac{\ln x}{x^5}\right]$$
$f(x) = \ln x$
$f'(x) = \frac{1}{x}$
$f(x) = x^5$
$f'(x) = 5x^4$

$$\frac{x^5\left(\frac{1}{x}\right) - \ln x(5x^4)}{(x^5)^2}$$

$$\frac{x^4(1 - 5\ln x)}{x^{10}}$$

9 IMPLICIT DIFFERENTIATION

$$3y^4 - 2x^2y = 3$$

$$\frac{d}{dx}[3y^4]$$

$$\frac{d}{dx}[3]$$

$$12y^3\frac{dy}{dx} - 2\left[x^2\frac{dy}{dx} + y(2x)\right] = 0$$

$$12y^3\frac{dy}{dx} - 2x^2\frac{dy}{dx} - 4xy = 0$$

$$(12y^3 - 2x^2)\frac{dy}{dx} = 4xy$$

$$\frac{dy}{dx}\left[\frac{4xy}{12y^3 - 2x^2}\right]$$

$$\frac{dy}{dx}\left[\frac{4xy}{2(6y^3 - x^2)}\right]$$

$$\frac{2xy}{6y^3 - x^2}$$

$$y^2 = e^{xy^3}$$

$$\frac{2y}{2y} = \frac{e^{xy^3}}{2y}$$

$$2y\frac{dy}{dx} = e^{xy^3}\left[x\left(3y^2\frac{dy}{dx}\right) + y^3\right]$$

$$2y\frac{dy}{dx} = 3xy^2 e^{xy^3}\frac{dy}{dx} + y^3 e^{xy^3}$$

$$2y\frac{dy}{dx} - 3xy^2 e^{xy^3}\frac{dy}{dx} = y^3 e^{xy^3}$$

$$\boxed{\frac{dy}{dx} = \frac{y^3 e^{xy^3}}{2y - 3xy^2 e^{xy^3}}}$$

10 RELATED RATES

$$y = \sqrt{x} \quad \frac{dy}{dt} \text{ when } x = 4 \quad \frac{dy}{dt} = 3$$

$$\frac{d}{dt}[y]$$
$$\frac{d}{dt}[\sqrt{x}]$$
$$\frac{d}{dt}\left[x^{\frac{1}{2}}\right]$$

$$\frac{dy}{dt} = \frac{1}{2} x^{-\frac{1}{2}} \frac{dx}{dt}$$

$$\frac{dy}{dt} = \frac{1}{2\sqrt{x}} \frac{dx}{dt}$$

$$\frac{dy}{dt} = \frac{1}{2\sqrt{4}}(3)$$

$$\boxed{\frac{dy}{dt} = \frac{3}{4}}$$

A ball is inflated with air at the rate of 800 cubic centimeters per minute. How fast is the radius of the ball increasing at the instant the radius is 30 centimeters?

$$v = \frac{4}{3}\pi r^3$$

$$\frac{dv}{dt}[v] \quad \frac{dv}{dt}\left[\frac{4}{3}\pi r^3\right]$$

$$\frac{dv}{dt} = \frac{4}{3}\pi\left(3r^2\frac{dr}{dt}\right)$$

$$\frac{dv}{dt} = 4\pi r^2 \frac{dr}{dt}$$

$$800 = 4\pi(30)^2 \frac{dr}{dt}$$

$$\frac{800}{3600\pi} = \frac{dr}{dt}$$

$$\boxed{\frac{2}{9\pi} = \frac{dr}{dt}}$$

A jet is flying at an altitude of 5 miles and passes directly over a weather station. When the jet is 10 miles away the station detects that the distance is changing at a rate of 240 miles per hour. What is the speed of the jet?

$5^2 + x^2 = s^2$

$25 + x^2 = s^2$

$\dfrac{2x}{2}\dfrac{dx}{dt} = \dfrac{2s}{2}\dfrac{ds}{dt}$

$x\dfrac{dx}{dt} = s\dfrac{ds}{dt}$

$\sqrt{75} = 5\sqrt{3}\dfrac{dx}{dt} = 10(240)$

$\dfrac{dx}{dt} = \dfrac{480}{\sqrt{3}}$

$\dfrac{480\sqrt{3}}{3}$

$\boxed{160\sqrt{3} \approx 277 mph}$

A rocket flies at an altitude of $y=8$ miles toward a point directly over an observer. The speed of the rocket is 400 miles per hour. Find the rates at which the angle of elevation θ is changing when the angle is $\theta = 45°, \theta = 60°, \theta = 80°$

$$\tan \theta = \frac{8}{x} \to x = 8 \cot \theta$$

$$\frac{dx}{dt} = 8\left(-\csc^2 \theta \frac{d\theta}{dt}\right)$$

$$\frac{dx}{dt} = -400$$

$$\frac{d\theta}{dt} = \left(\frac{400}{8(\csc^2 \theta)}\right) rad/h$$

$$\frac{400}{8} = 50 \sin^2 \theta$$

When $\theta = 45° \to \frac{\pi}{4}$

When $\theta = 60° \rightarrow \frac{\pi}{3}$

When $\theta = 80° \rightarrow \frac{4\pi}{9}$

$50(\sin 45°)^2 = 25 \div 60 (min)$

$\boxed{\dfrac{5}{12}}$

$50(\sin 60°)^2 = \dfrac{75}{2} \div 60 (min)$

$\boxed{\dfrac{5}{8}}$

$50(\sin 80°)^2 = 48.49.. \div 60 (min)$

$\boxed{0.808 \ldots \approx 0.81}$

11 Extrema on an Interval

$f(x) = 2x^3 - 6x, [0,3]$

$f'(x) = 6x^2 - 6$

$0 = 6(x+1)(x-1)$

~~$x + 1 = 0$~~ $x - 1 = 0$
~~$x = -1$~~ $x = 1$

$f(1) = 2(1)^3 - 6(1) = -4$
$f(0) = 2(0)^3 - 6(0) = 0$
$f(0) = 2(3)^3 - 6(3) = 36$
$Minimum\ (1, -4)$
$Maximum\ (3, 36)$

12 INCREASING AND DECREASING FUNCTIONS AND THE FIRST DERIVATIVE TEST

$f(x) = x^3 - 6x^2 + 15$

(a) $\quad f'(x) = 3x^2 - 12x$
$\quad\quad 0 = 3x^2 - 12x$
$\quad\quad\quad 3x(x-4)$

Critical values
$3x = 0 \quad x - 4 = 0$
$x = 0 \quad\quad x = 4$

(b) $\quad (-\infty, 0) \quad x = -1 \quad f'(-1) = (+)$
$\quad\quad (0, 4) \quad\quad x = 1 \quad\quad f'(1) = (-)$
$\quad\quad (4, \infty) \quad\quad x = 5 \quad\quad f'(5) = (+)$

Increasing $(-\infty, 0)$ and $(4, \infty)$
$\quad$ Decreasing $(0, 4)$
$\quad$ Relative max $(0, 15)$
$\quad$ Relative min $(4, -17)$

$f(x) = \sin x \cos x + 5$
$f'(x) = \cos^2 x - \sin^2 x$

$$0 = (1 - \sin^2 x) - \sin^2 x$$
$$0 = 1 - 2\sin^2 x$$
$$\sin^2 x = \frac{1}{2}$$
$$\sin x = \pm\sqrt{\frac{1}{2}} \rightarrow \pm\sqrt{\frac{2}{2}}$$

$\sin x = \pm\sqrt{\frac{2}{2}} \, (45°) Unit\ Circle$

$\frac{\pi}{4}, \frac{3\pi}{4}, \frac{5\pi}{4}, \frac{7\pi}{4} \left(Add \frac{\pi}{2}\right)$

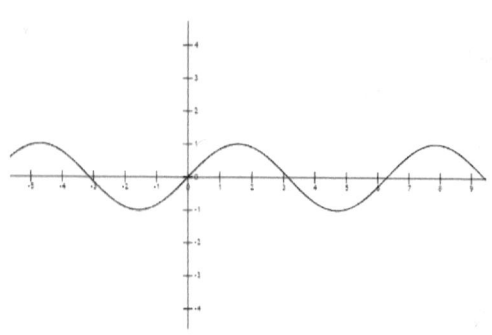

$$\left(0, \frac{\pi}{4}\right), \left(\frac{\pi}{4}, \frac{3\pi}{4}\right), \left(\frac{3\pi}{4}, \frac{5\pi}{4}\right), \left(\frac{5\pi}{4}, \frac{7\pi}{4}\right), \left(\frac{7\pi}{4}, 2\pi\right)$$

Increase $\left(0, \frac{\pi}{4}\right), \left(\frac{3\pi}{4}, \frac{5\pi}{4}\right), \left(\frac{7\pi}{4}, 2\pi\right)$

Decrease $\left(\frac{\pi}{4}, \frac{3\pi}{4}\right), \left(\frac{5\pi}{4}, \frac{7\pi}{4}\right)$

Relative max $\frac{\pi}{4}, \frac{5\pi}{4}$

Relative min $\frac{3\pi}{4}, \frac{7\pi}{4}$

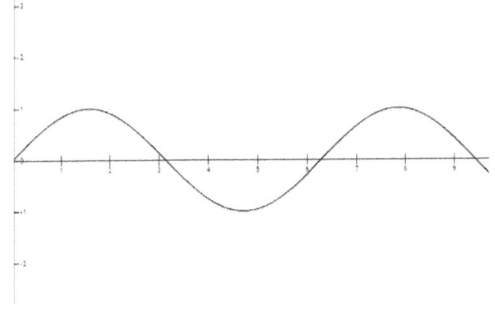

13 CONCAVITY AND SECOND DERIVATIVE TEST

$$f(x) = \frac{2x^2}{3x^2 + 1}$$

$f'(x) = 2x^2 \rightarrow 4x$
$f'(x) = 3x^2 + 1 \rightarrow 6x$

$$\frac{3x^2 + 1(4x) - 2x^2(6x)}{(3x^2 + 1)^2}$$

$$f'(x) = \frac{4x}{(3x^2 + 1)^2}$$

Second Derivative:
$$\frac{3x^2 + 1(4) - 4x[2(3x^2 + 1)6x]}{(3x^2 + 1)^4}$$

$$\frac{4(3x^2 + 1)[3x^2 + 1 - 12x^2]}{(3x^2 + 1)^4}$$

$$f''(x) = \frac{4(1 - 9x^2)}{(3x^2 + 1)^3}$$

$4(1 - 9x^2) = 0$

$$x = \pm \frac{1}{3}$$

$\left(-\infty, -\frac{1}{3}\right), \left(-\frac{1}{3}, \frac{1}{3}\right), \left(\frac{1}{3}, \infty\right)$

$f''(-1) = (-) \rightarrow$ Concave down
$f''(0) = (+) \rightarrow$ Concave up
$f''(1) = (-) \rightarrow$ Concave down

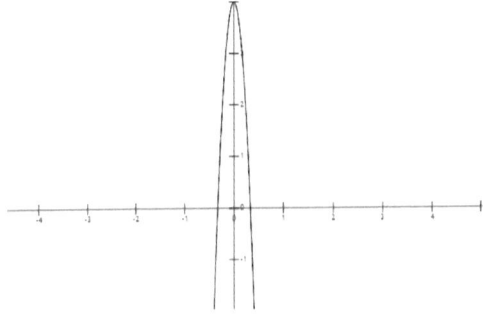

14 OPTIMIZATION

The sum is s and the product is a maximum.
$$f(x,y) = xy \text{ (Primary)}$$
$$x + y = s \text{ (Constraint equation)}$$

$$y = s - x$$

$$f(x, s-x) = x(s-x)$$

$$f(x) = sx - x^2$$
$$f'(x) = s - 2x$$
$$f''(x) = -2$$

$$\boxed{x = \frac{s}{2}}$$

The product is 185 and the sum is a minimum
$$f(x,y) = x + y$$
$$xy = 185$$

$$f(x) = x + \frac{185}{x}$$
$$f'(x) = 1 - \frac{185}{x^2}$$

$$x^2 = 185$$
$$\sqrt{185}$$

$$f''(x) = \frac{370}{x^3}$$

$$f''(\sqrt{185}) = \frac{370}{\left(\sqrt{185}^3\right)} \to (+)$$

$$\boxed{x = \sqrt{185}}$$

Find the length and width of a rectangle that has the given perimeter of 80 m.

$$A(\ell, w) = \ell w \ (Primary)$$
$$80 = 2\ell + 2w \ (Secondary)$$

$$A(w) = (40 - w)w$$
$$A'(w) = 40 - 2w$$

$$0 = 40 - 2w$$
$$\boxed{w = 20, \ell = 20}$$

Find the length and width of a rectangle that has the given a maximum area of 32 square feet.

$$p(\ell, w) = 2\ell + 2w$$
$$32 = \ell w$$

$$p(w) = 2\left(\frac{32}{w}\right) + 2w$$
$$p(w) = \left(\frac{64}{w}\right) + 2w$$

$$\frac{-64}{w^2} + 2$$

$$\boxed{w = \sqrt{32} \to 4\sqrt{2} \quad \ell = \sqrt{32} \to 4\sqrt{2}}$$

15 DIFFERENTIAL

Function	x – value	Differential of x
		$\Delta x = dx = 0.1$
		$dy = f'(x)dx$
$y = 6 - 2x^2$		$\Delta y = y_2 - y_1$
$(-2, -2)$	$x = -1$	$\Delta y = .78$
$(-1.9, -1.22)$		$y' = -4x$
		$f'(2) = -4(-2) = 8$
		$8(0.1) = .8$

The measurement of the radius of a circle is 16 inches, with a possible error of $\frac{1}{4}$ inch. Use differentials to find the possible error in finding the area of the circle

$A = \pi r^2$

$\dfrac{dA}{dr} = 2\pi r\, dr$

$dA = 2\pi(16)\left(\dfrac{1}{4}\right) = \boxed{8\pi}$

16 ANTI-DERIVATIVES AND INDEFINITE INTEGRATION

$$\int 3x^4 \, dx$$

$$\int \frac{1}{\sqrt[4]{x^3}} \, dx$$

$$3\left[\frac{x^{4+1}}{4+1}\right]$$

$$\int x^{-\frac{3}{4}} \, dx$$

$$\boxed{\frac{3}{5}x^5 + C}$$

$$\frac{x^{\frac{1}{4}}}{\frac{1}{4}}$$

$$\int \frac{1}{x} \, dx$$

$$\boxed{4x^{\frac{1}{4}} + C}$$

$$\int x^{-1} \, dx$$

$$\boxed{\ln|x| + C}$$

17 INTEGRATION BY SUBSTITUTION

$$\int \sin^4 \theta \cos \theta \, d\theta \qquad \int (3x^2 + 2)^4 6x \, dx$$

$u = \sin \theta \quad du = \cos \theta \, d\theta$

$u = 3x^2 + 2 \qquad \dfrac{du}{dx} = 6x$
$\qquad\qquad\qquad\quad du = 6x\,dx$

$\int (\sin \theta)^4 \cos \theta \, d\theta$

$\int u^4 \, du \to \dfrac{u^{4+1}}{5} = \dfrac{u^5}{5}$

$\int u^4 \, du \to \dfrac{u^{4+1}}{5} = \dfrac{u^5}{5}$

$$\boxed{\dfrac{(3x^2 + 2)^5}{5} + C}$$

$$\boxed{\dfrac{\sin^5 \theta}{5} + C}$$

$$\int x\sqrt{x^2-5}\,dx$$

$$\boxed{\frac{1}{3}(x^2-5)^{\frac{3}{2}}+C}$$

$$u = x^2 - 5 \quad du = 2x\,dx$$

$$\int \tan\theta\,d\theta$$

$$\frac{1}{2}\int (x^2-5)^{\frac{1}{2}} 2x\,dx$$

$$-\int \frac{-\sin\theta}{\cos\theta}$$

$$\int u^{\frac{1}{2}} 2x\,dx$$
$$\frac{1}{2}\int u^{\frac{1}{2}}\,dx$$

$$u = \cos\theta \quad du = -\sin\theta\,d\theta$$

$$-\int \frac{1}{u}\,du$$

$$\frac{1}{2}\left[\frac{u^{\frac{3}{2}}}{\frac{3}{2}} + C\right]$$

$$-\ln|u| + C$$

$$\frac{1}{2}\left[\frac{2}{3}u^{\frac{3}{2}} + C\right]$$

$$\boxed{-\ln|\cos\theta| + C}$$

$$\frac{1}{3}u^{\frac{3}{2}} + C$$

$$\int \frac{x^3 - 8x}{x^2} dx$$

$$\int \frac{x(x^2 - 8)}{x^2} dx$$

$$\int \frac{(x^2 - 8)}{x}$$

$$\int \frac{x^2}{x} - \frac{8}{x}$$

$$\int x - \frac{8}{x}$$

$$\int x\,dx - 8 \int \frac{1}{x} dx$$

$$\boxed{\frac{x^2}{2} - 8\ln|x| + C}$$

18
RIEMANN SUMS & DEFINITE INTEGRALS

Divide the Interval [0, 4] into four subinterval

$$\frac{4-0}{4} = 1$$

[0,1], [1,2], [2,3], [3,4]

Divide the Interval [0, 4] into five subinterval

$$\frac{4-0}{5} = \frac{4}{5}$$

$$\left[0, \frac{4}{5}\right], \left[\frac{4}{5}, \frac{8}{5}\right], \left[\frac{8}{5}, \frac{12}{5}\right], \left[\frac{12}{5}, \frac{16}{5}\right], \left[\frac{16}{5}, 4\right]$$

Divide the Interval [2, 5] into five subinterval

$$\frac{5-2}{5} = \frac{3}{5}$$

$$\left[2, \frac{13}{5}\right], \left[\frac{13}{5}, \frac{16}{5}\right], \left[\frac{16}{5}, \frac{19}{5}\right], \left[\frac{19}{5}, \frac{22}{5}\right], \left[\frac{22}{5}, 5\right]$$

Find the area between the graph of $f(x) = x^2 + 1$ and the x-axis using n inscribed rectangles

$$\Delta x = \frac{2-0}{n}$$

$$\underbrace{\left[0, \frac{2}{n}\right]}_{1^{st}} \underbrace{\left[\frac{2}{n}, \frac{4}{n}\right]}_{2^{nd}} \underbrace{\left[\frac{4}{n}, \frac{6}{n}\right]}_{3^{rd}} \ldots \underbrace{\left[(i-1)\frac{2}{n}, i\frac{2}{n}\right]}_{i^{th}} \underbrace{\left[(n-1)\frac{2}{n}, 2\right]}_{n^{th}}$$

$$\frac{2}{n}[(0^2)+1] + \frac{2}{n}\left[\left(\frac{2}{n}\right)^2 + 1\right] + \frac{2}{n}\left[\left(\frac{4}{n}\right)^2 + 1\right]$$
$$+ \frac{2}{n}\left[\left(\frac{6}{n}\right)^2 + 1\right] + \frac{2}{n}\left[\left((i-1)\frac{2}{n}\right)^2 + 1\right]$$
$$+ \frac{2}{n}\left[\left((n-1)\frac{2}{n}\right)^2 + 1\right]$$

$$\sum_{i=1}^{n} \frac{2}{n}\left[\left((i-1)\frac{2}{n}\right)^2 + 1\right]$$

$$\frac{2}{n}\sum_{i=1}^{n}\left[\left((i-1)\frac{2}{n}\right)^2 + 1\right]$$

$$\frac{2}{n}\left[\sum_{i=1}^{n}\left[\left((i-1)\frac{2}{n}\right)^2 + \sum_{i=1}^{n} 1\right]\right]$$

$$\frac{2}{n}\left[\sum_{i=1}^{n}\left[\left((i-1)\frac{2}{n}\right)^2 + n\right]\right]$$

$$\frac{2}{n}\left[\sum_{i=1}^{n}\left[(i^2 - 2i + 1)\frac{4}{n^2}\right] + n\right]$$

$$\frac{2}{n}\left[\frac{4}{n^2}\sum_{i=1}^{n} i^2 - \frac{8}{n^2}\sum_{i=0}^{n} + \frac{4}{n^2}\sum_{i=1}^{n} 1 + n\right]$$

Approximation area:

$$\frac{2}{n}\left[\frac{4^2}{n^2}\left(\frac{n(n+1)(2n+1)}{6}\right) - \frac{8}{n^2}\left(\frac{n}{2}(n+1)\right) + \frac{4}{n^2}(n) + n\right]$$

Exact area:

$$\lim_{n \to \infty} \left(\frac{2}{n} \left[\frac{2(n+1)(2n+1)}{3n} - \frac{4(n+1)}{n} + \frac{4}{n} + n \right] \right)$$

$$\lim_{n \to \infty} \left(\left[\frac{4(n+1)(2n+1)}{3n^2} - \frac{8(n+1)}{n^2} + \frac{4}{n^2} \right] \right)$$

$$\frac{8}{3} - 0 - 0 + 2 = \boxed{\frac{14}{3}}$$

Find the area of the region under the graphs of

$x=a$, $x=b$, $f(x)$ and the x-axis

Regular Partition:

$$\Delta x = \frac{b-a}{n}$$

i^{th} Interval

$$[a, a + \Delta x][a + \Delta x, a + 2\Delta x][a + 2\Delta x, a + 3\Delta x] \ldots [a + (i-1)\Delta x, a + i\Delta x]$$

$$\left[a + (i-1)\left(\frac{b-a}{n}\right), a + i\left(\frac{b-a}{n}\right)\right]$$

$$\frac{b-a}{n} f\left(a + i\left(\frac{b-a}{n}\right)\right)$$

Approximate Area:

$$\boxed{\sum_{i=1}^{n} \frac{b-a}{n} f\left(a + i\left(\frac{b-a}{n}\right)\right)}$$

$$\int_{1}^{2} 2x^2 \sqrt{x^3 + 1}\, dx$$

$$u = x^3 + 1 \quad du = 3x^2\, dx$$

$$\frac{2}{3} \int_{1}^{2} \left(\sqrt{x^3 + 1}\right)(3x^2\, dx)$$

$$\frac{2}{3} \int_{1}^{2} \left(\sqrt{x^3 + 1}\right)(3x^2\, dx)$$

$$\frac{2}{3}\int_{2}^{9} u^{\frac{1}{2}}\,du$$

$$\frac{2}{3}\left[\frac{2}{3}u^{\frac{3}{2}}\right]_{2}^{9}$$

$$\frac{4}{9}\left[9^{\frac{3}{2}} - 2^{\frac{3}{2}}\right]$$

$$\frac{4}{9}\left(27 - 2\sqrt{2}\right)$$

$$\boxed{12 - \frac{8\sqrt{2}}{9}}$$

$f(x) = 2x^2 + 3, [-2,1]$

$\int_{-2}^{1} (2x^2 + 3)\, dx$

$\Delta x = \dfrac{b-a}{n} = \dfrac{3}{n}$

$[a + (i-1)\Delta x, a + i\Delta x]$

$\left[-2 + (i-1)\dfrac{3}{n}, -2 + i\dfrac{3}{n}\right]$

$c_i = -2 + i\dfrac{3}{n}$

$\displaystyle\lim_{n\to\infty} \sum_{i=1}^{n} f(c_i)\Delta x$

$\displaystyle\lim_{n\to\infty} \sum_{i=1}^{n} \left[2\left(-2 + \dfrac{3i}{n}\right)^2 + 3\right]\dfrac{3}{n}$

$$\lim_{n\to\infty}\frac{3}{n}\sum_{i=0}^{n}\left[2\left(4-\frac{12i}{n}+\frac{9i^2}{n}\right)+3\right]$$

$$\frac{11}{3}-\left(-\frac{34}{3}\right)=\boxed{15}$$

The end

Bibliography:

Ocal. Soccer Ball Clip Art. Digital image. ClipShrine.com. N.p., 26 June 2012. Web. 17 Feb. 2017. <http://www.clipshrine.com/Soccer-Ball-5587-medium.html>.

Wetzel, John. Fifties Jet for Kinematics Illustrations. Digital image. The WikiPremed. N.p., n.d. Web. 17 Jan. 2017. <http://www.wikipremed.com/image.php?img=010101_68zzzz100350_00104_68.jpg&image_id=100350>.

www.ingramcontent.com/pod-product-compliance
Lightning Source LLC
Chambersburg PA
CBHW020710180526
45163CB00008B/3028